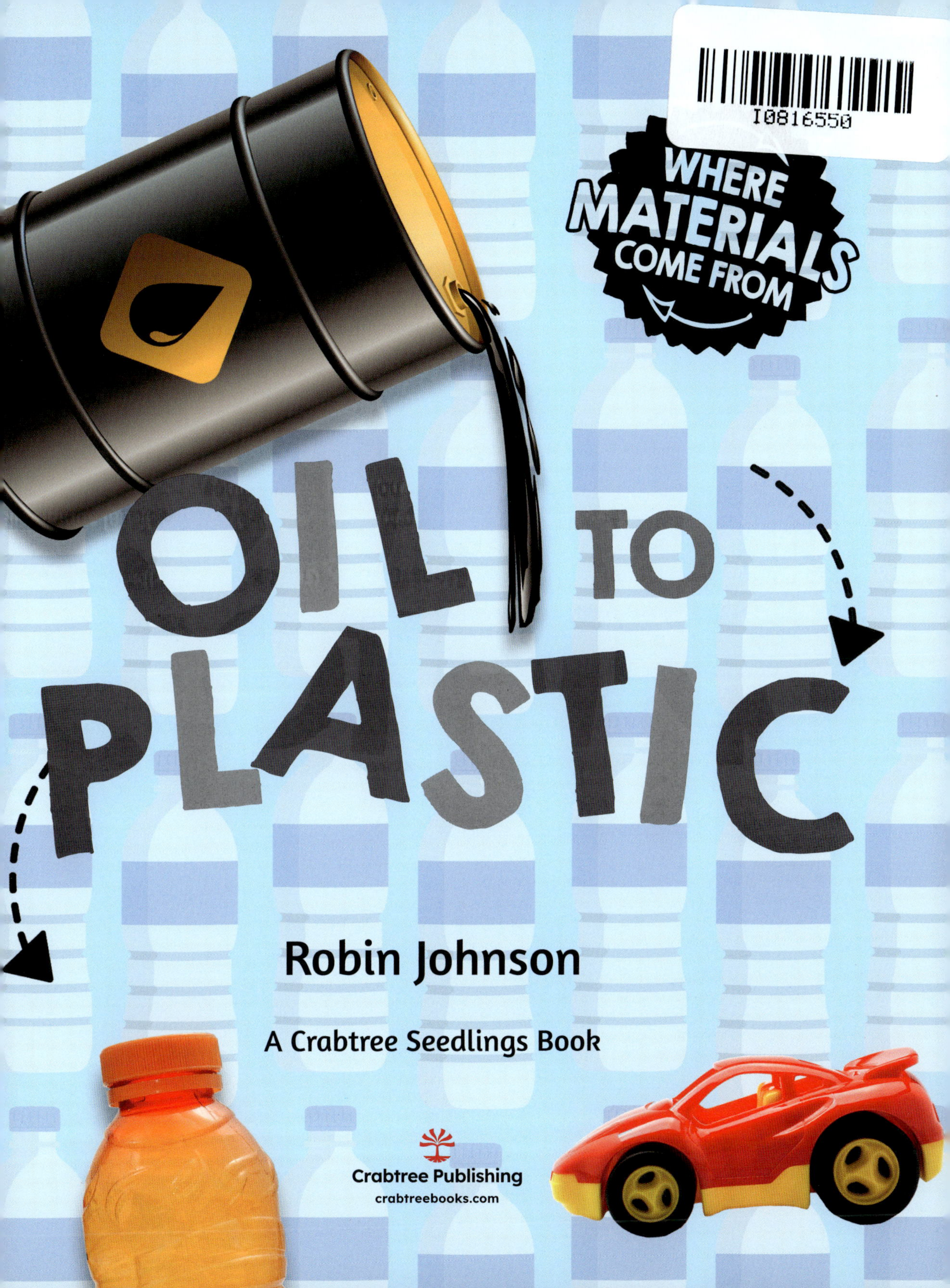

OIL TO PLASTIC

Robin Johnson

A Crabtree Seedlings Book

Crabtree Publishing
crabtreebooks.com

Author: Robin Johnson

Editor: Ellen Rodger

Proofreader: Melissa Boyce

Design: Katherine Berti

Photo research: Robin Johnson, Katherine Berti

Prepress and print coordinator: Katherine Berti

Photographs and illustrations:

Shutterstock
- Lakeview Images: p. 8 (right)
- Larina Marina: p. 21 (top left)
- pim pic: p. 23 (bottom)
- Vladimir Endovitskiy: p. 10

All other images from Shutterstock

Crabtree Publishing

crabtreebooks.com 800-387-7650

In Canada: We acknowledge the financial support of the Government of Canada through the Canada Book Fund for our publishing activities.

Hardcover 978-1-0398-0660-3
Paperback 978-1-0398-0686-3
Ebook (pdf) 978-1-0398-0710-5
Epub 978-1-0398-0737-2

Published in Canada
Crabtree Publishing
616 Welland Avenue
St. Catharines, Ontario
L2M 5V6

Published in the United States
Crabtree Publishing
347 Fifth Avenue
Suite 1402-145
New York, New York, 10016

Library and Archives Canada Cataloguing in Publication
Available at Library and Archives Canada

Library of Congress Cataloging-in-Publication Data
Available at the Library of Congress

Printed in the U.S.A./012023/CG20220815

Contents

Do you play with plastic toys? Plastic is a material used in many things. We use these things every day. But what is plastic? Where does it come from?

What Is Plastic?

Plastic is a material made by people. It can be formed into almost any shape. Plastic is useful because it is strong, lightweight, and lasts for a long time.

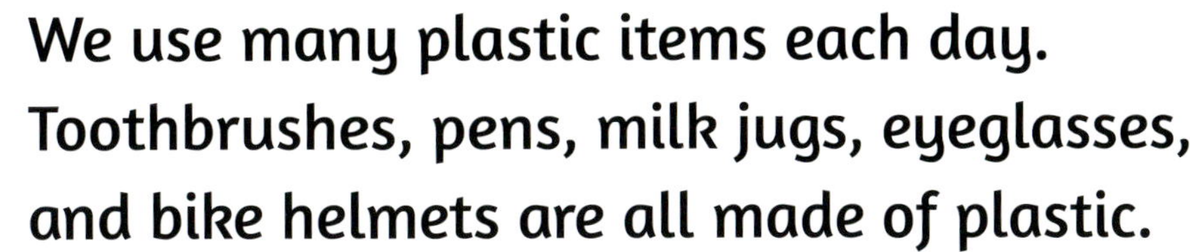

We use many plastic items each day. Toothbrushes, pens, milk jugs, eyeglasses, and bike helmets are all made of plastic.

PLASTIC FACT

There are many kinds of plastic. Some are hard and tough. Others are soft and stretchy.

From Fossil Fuels

Most plastic is made from **fossil fuels**. Fossil fuels formed in the earth over millions of years. These resources cannot be replaced once they are used up.

Crude oil, natural gas, and coal are types of fossil fuels. They can all be used to make plastic. In this book, you will learn how plastic is made from crude oil.

PLASTIC FACT

Some newer plastics are made from corn, potatoes, and other plants.

Finding Oil

Crude oil is found deep beneath the ground and ocean floor. It collects in areas of rock called reservoirs. These rocks have billions of tiny holes or larger bowl-shaped spaces that hold oil.

PLASTIC FACT

Crude oil sometimes rises to the surface. This is called an oil seep.

Scientists try to **predict** where oil is located. They study rocks above and below the surface. They make maps and do many tests.

Well Drilling

When scientists predict there is oil in an area, people drill deep wells to explore under the ground. Scientists study the rocks, mud, and liquids in the wells.

If there is a good supply of oil, workers remove the oil from the ground. They join long steel pipes together and lower them into the wells. Then they pump the oil up the pipes to the surface.

WONDER WORD:
WELL

A well is a deep hole made in the ground to reach oil, water, or other natural resources.

Transporting Oil

Crude oil is sent to different places to be stored or cleaned. Most of the oil is pumped through a long system of pipes called a pipeline.

Crude oil is also shipped over land by trains and trucks. It is transported on water by huge ships called tankers.

PLASTIC FACT

Sometimes oil can spill when it is being shipped. Oil spills can kill plants and animals and hurt the **environment**.

Oil Refineries

Crude oil is shipped to huge factories called refineries. There, machines heat and cool the oil. These processes clean the oil and separate it into small parts called **molecules**.

Scientists join the molecules together in different ways. Joined molecules are called polymers. They are melted, cooled, and cut into small plastic **pellets.**

WONDER WORD:
POLYMER

Polymer is a Greek word that means many parts.

Taking Shape

Plastic pellets are shipped to factories. There, people melt the pellets down to a thick liquid. Then they pour the liquid plastic into **molds**.

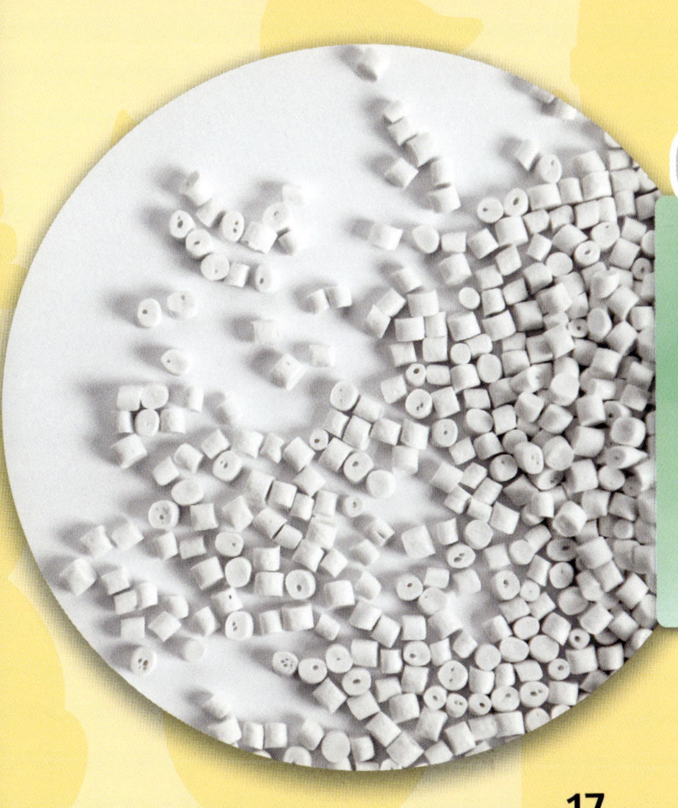

PLASTIC FACT

Plastic pellets are naturally clear or milky white. People add substances to them to change their color.

Plastic Products

Many **industries** use plastic to make products or do work. The auto industry uses plastic to make car doors, seats, and other parts and to test that cars are safe.

Construction workers use pipes and other plastic items to build homes. Hard plastic hats keep them safe at work.

A lot of plastic is used to package food and other products. Plastic packaging keeps food clean and fresh.

Plastic Problems

Plastic is a very useful material. But making plastic can harm the environment. It uses up fossil fuels. Refineries and factories create pollution. Plastic pellets sometimes spill and get washed into the ocean.

Plastic does not break down easily, so it piles up in **landfills**. Plastic bags, bottles, and other items often end up in the ocean. Sea animals can get hurt by the plastic.

WONDER WORD:
POLLUTION

Pollution is harmful materials that can make the air, water, and soil unclean.

How Can You Help?

You can help the environment by using less plastic. Avoid using plastics that get thrown away after one use, such as straws or grocery bags. Choose items that can be reused instead.

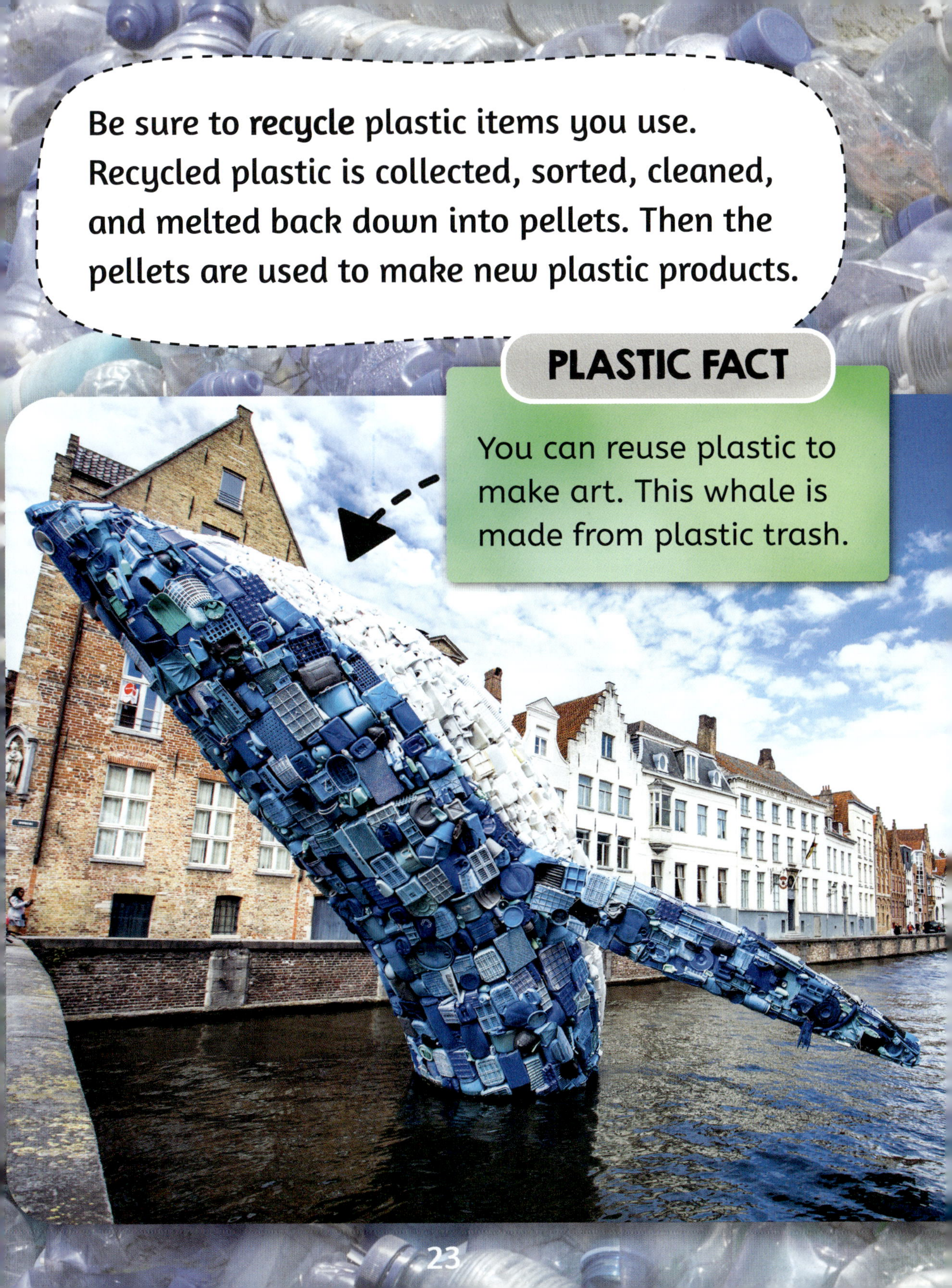

Be sure to **recycle** plastic items you use. Recycled plastic is collected, sorted, cleaned, and melted back down into pellets. Then the pellets are used to make new plastic products.

PLASTIC FACT

You can reuse plastic to make art. This whale is made from plastic trash.

GLOSSARY

crude oil A thick, black liquid found in nature that people use to make plastic, gasoline, and other products

environment All the physical surroundings on Earth, including the air, soil, water, plants, and animals

fossil fuels Substances such as oil and coal that formed in the ground over millions of years from the remains of tiny plants and animals

industries Groups of businesses that make or sell similar products or do similar jobs

landfill A place where garbage is taken and buried in the ground; often called a dump

molds Hollow forms in which something is shaped

molecules The smallest portions of a substance that have all the properties of that substance

pellets Small round or tube-shaped pieces of plastic or other material

predict To tell what will happen before it takes place

recycle To change trash into materials that can be used again

INDEX